THIS BOOK BELONGS TO:

_____

The Wonderful World of Space

Mimi Jones

Dedicated to my fourth child, Mars.

All rights reserved.
No part of this book may be reproduced in any form or by any means, electronic or mechanical, and no photocopying or recording, unless you have written permission from the author.

ISBN 978-1-958985-34-2

Text copyright © 2024 by Mimi Jones

www.joeysavestheday.com

A Mimi Book

WELCOME TO THE WONDERFUL WORLD OF SPACE

Our solar system is a captivating and diverse assembly of celestial objects orbiting the Sun, a star at its core. It comprises eight primary planets: the rocky inner planets Mercury, Venus, Earth, and Mars, alongside the gas giants Jupiter and Saturn, and the ice giants Uranus and Neptune. Besides these planets, the solar system hosts dwarf planets like Pluto, numerous moons, and countless asteroids and comets. The Kuiper Belt and the Oort Cloud, regions teeming with icy bodies, extend well beyond Neptune's orbit. As a part of the Milky Way galaxy, our solar system continues to enchant scientists and stargazers with its intricate complexity and stunning beauty.

The Sun, the heart of our solar system, is predominantly composed of hydrogen and helium. It generates light and heat through nuclear fusion, which is crucial for sustaining life on Earth. Enormously larger than Earth, the Sun contains over 99% of the solar system's mass. Its powerful gravity holds planets, asteroids, and comets in their orbits. The Sun's surface reaches scorching temperatures of about 5,500 degrees Celsius (9,932 degrees Fahrenheit), while its core is even more blistering, at around 15 million degrees Celsius (27 million degrees Fahrenheit).

PLANET MERCURY

Mercury is the closest planet to the Sun and the smallest in our solar system. It experiences extreme temperature variations, with scorching daytime highs and freezing nighttime lows. Its cratered surface resembles our Moon, and it has a thin atmosphere composed mainly of oxygen, sodium, and hydrogen. Mercury's year lasts 88 Earth days, but one day-night cycle takes 176 Earth days due to its slow rotation.

MERCURY
SOLAR SYSTEM

Venus, the second planet from the Sun, is similar in size and composition to Earth but has a dense atmosphere of carbon dioxide and sulfuric acid clouds. This creates a severe greenhouse effect, making its surface extremely hot. Venus has a landscape of plains, volcanoes, and canyons. It rotates slowly and in the opposite direction of its orbit, causing a single day to last longer than its year.

PLANET VENUS

Earth is the third planet from the Sun and is unique for its ability to support life. It has a variety of ecosystems and climates, and its atmosphere is rich in nitrogen and oxygen. The dynamic surface is made up of tectonic plates, causing earthquakes and volcanic activity. Home to millions of species, including humans, Earth maintains a delicate balance of life and natural processes, making it a precious part of our solar system.

PLANET EARTH

PLANET MARS

Mars, the fourth planet from the Sun, is known as the "Red Planet" because of its iron oxide surface. It features the tallest volcano, Olympus Mons, and the deepest canyon, Valles Marineris, in the solar system. Despite its thin atmosphere of carbon dioxide, Mars has polar ice caps made of water and dry ice. Ancient riverbeds and lake signs indicate a wetter past. Mars is a key focus for exploration, seeking signs of life and potential human colonization.

Jupiter, the fifth planet from the Sun, is the largest in our solar system. Known for its colorful bands and the Great Red Spot, a massive storm, Jupiter's atmosphere is mainly hydrogen and helium. It has a strong magnetic field, a faint ring system, and numerous moons, including Ganymede, Io, Europa, and Callisto, each with unique features. Jupiter's size and dynamic nature provide insights into our solar system's formation and evolution.

PLANET JUPITER

Saturn, the sixth planet from the Sun, is renowned for its beautiful and complex ring system made of ice, rock, and dust. This gas giant, primarily composed of hydrogen and helium, experiences strong winds and atmospheric storms. With over 80 moons, including the intriguing Titan, Saturn offers unique features that captivate scientists and space enthusiasts, providing valuable insights into the complexities of our solar system.

PLANET SATURN

Uranus, the seventh planet from the Sun, is an ice giant with a unique sideways rotation that causes extreme seasonal variations. Its atmosphere consists mainly of hydrogen, helium, and methane, giving it a blue-green color. Uranus has a faint ring system and at least 27 moons, with Titania and Oberon being the largest. Its unusual characteristics and distant, cold environment make Uranus a fascinating subject of study.

PLANET URANUS

Neptune, the eighth and farthest planet from the Sun, is an ice giant known for its deep blue color due to methane in its atmosphere. It has the fastest winds in the solar system and dynamic weather, including the Great Dark Spot storm. Neptune has a faint ring system and at least 14 moons, with Triton being the largest and most intriguing. Despite its distance, Neptune continues to fascinate scientists with insights into the outer reaches of our solar system.

PLANET NEPTUNE

Ceres, the largest object in the asteroid belt between Mars and Jupiter, is a dwarf planet with a diameter of about 940 kilometers (585 miles). It has a rocky core, an icy mantle, and a surface with craters and bright spots, likely salt or ice deposits. Water vapor detected by space missions suggests subsurface water, making Ceres a key interest in the search for extraterrestrial life.

Dwarf Planet Ceres

Pluto, once the ninth planet, is now classified as a dwarf planet and resides in the Kuiper Belt. It has a diameter of about 2,377 kilometers (1,477 miles) and is made of ice and rock. Pluto has a thin, seasonally changing atmosphere and a surface with mountains, valleys, and nitrogen ice plains. Its largest moon, Charon, is nearly half its size. The New Horizons mission in 2015 revealed Pluto to be a complex and geologically active world.

Dwarf Planet Pluto

Haumea is a dwarf planet in the Kuiper Belt, known for its elongated, ellipsoid shape due to its rapid rotation, which completes a rotation every four hours. Its surface is covered in crystalline water ice, and it has two moons, Hi'iaka and Namaka. Haumea's unique characteristics make it an intriguing subject for astronomers studying the diversity of objects in our solar system.

Dwarf Planet Haumea

Makemake is a dwarf planet in the Kuiper Belt, named after the Rapa Nui god of fertility. It is the second-brightest object in the Kuiper Belt after Pluto and has a diameter of about 1,430 kilometers (890 miles). Makemake's surface is covered in methane, ethane, and possibly nitrogen ices, giving it a reddish-brown hue. Unlike many other dwarf planets, Makemake has no known moons, making it unique.

Dwarf Planet Makemake

Eris is a distant dwarf planet located in the scattered disk region beyond the Kuiper Belt. It is slightly smaller but more massive than Pluto, with a diameter of about 2,326 kilometers (1,445 miles) and an icy surface. Eris has one known moon, Dysnomia. Its discovery in 2005 played a key role in redefining what constitutes a planet in our solar system.

Dwarf Planet Eris

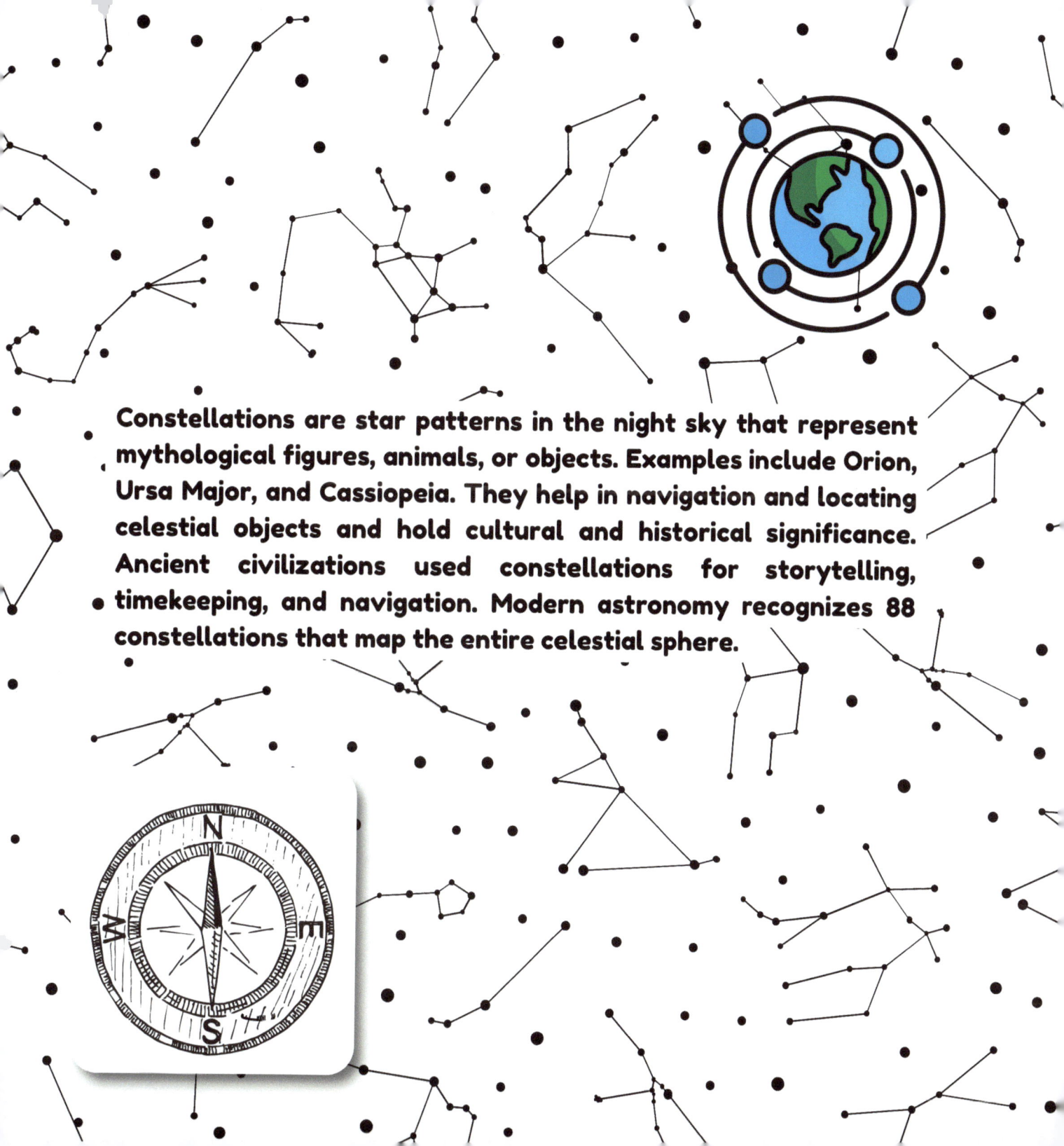

Constellations are star patterns in the night sky that represent mythological figures, animals, or objects. Examples include Orion, Ursa Major, and Cassiopeia. They help in navigation and locating celestial objects and hold cultural and historical significance. Ancient civilizations used constellations for storytelling, timekeeping, and navigation. Modern astronomy recognizes 88 constellations that map the entire celestial sphere.

A galaxy is a vast collection of stars, stellar remnants, gas, dust, and dark matter, bound by gravity. The Milky Way is our home galaxy, a barred spiral with billions of stars, including the Sun. Galaxies vary in shape and size and form clusters and superclusters, creating large-scale structures in the universe. Studying galaxies reveals insights into the cosmos' formation and evolution over billions of years.

Stars are luminous spheres of plasma held together by their own gravity. They vary in size, mass, temperature, and color, and they produce light and heat through the process of nuclear fusion, where hydrogen is converted into helium in their cores. The life cycle of a star includes several stages: formation, the main sequence, and eventual death. This final stage can result in phenomena such as supernovas, black holes, or white dwarfs. For millennia, stars have captivated humanity, serving as navigational aids, sources of inspiration, and subjects of scientific study.

The Moon, Earth's only natural satellite, has a diameter of about 3,474 kilometers (2,159 miles) and features geological structures like maria, craters, and mountain ranges. Its phases result from changing angles of sunlight as it orbits Earth, influencing timekeeping and cultural traditions. The Moon's gravitational pull affects ocean tides and has influenced Earth's evolution. Human exploration, particularly the Apollo missions, has revealed insights into its composition and the early solar system, making it a key subject of scientific study.

Black holes are regions of space with gravity so strong that nothing, not even light, can escape. They form from the remnants of massive stars after gravitational collapse and are defined by their event horizon, beyond which nothing can return. Black holes range from stellar-mass to supermassive, found at the centers of galaxies. Studying them provides insights into space, time, and the universe's mysteries.

The Milky Way galaxy, our cosmic home, is a barred spiral galaxy that spans about 100,000 light-years and contains billions of stars, including our Sun. Positioned within the Orion Arm, the Solar System orbits the galaxy's center at about 27,000 light-years. The Milky Way is known for its spiral arms, star formation regions, nebulae, and interstellar dust, with a supermassive black hole, Sagittarius A*, at its core. It also has diverse celestial objects, including star clusters, globular clusters, planets, and moons. As part of the Local Group, it interacts with neighboring galaxies like Andromeda, offering insights into galaxy structure and evolution.

Asteroids, comets, and meteors are intriguing celestial objects in our solar system. Asteroids are rocky remnants mainly found in the asteroid belt between Mars and Jupiter. Comets, composed of ice and dust, originate from the outer solar system and develop glowing comas and tails when near the Sun. Meteors, or "shooting stars," are the streaks of light from meteoroids burning up in Earth's atmosphere, and any surviving fragments that reach the surface are called meteorites. Studying these objects helps us understand the building blocks and formation processes of the solar system.

Solar winds are streams of charged particles, mainly electrons and protons, that flow from the Sun's corona through the solar system. These winds interact with planetary magnetospheres and atmospheres, creating phenomena like auroras. They also form the heliosphere, a protective bubble extending beyond Pluto, shielding against cosmic radiation. Understanding solar winds is essential for predicting space weather and mitigating their effects on satellite operations, communication systems, and power grids.

Solar flares are intense bursts of radiation and energy from the Sun's surface, usually near sunspots and active regions. They result from the sudden release of magnetic energy and emit radiation across the electromagnetic spectrum. Solar flares can impact space weather, affecting satellite communications, navigation systems, and power grids on Earth. They also accelerate particles that can interact with Earth's magnetic field, enhancing auroras and sometimes disrupting technological systems. Understanding solar flares is essential for predicting and mitigating their effects on our technology-dependent society.

The Kuiper Belt is a vast, donut-shaped region beyond Neptune's orbit, spanning approximately 30 to 55 astronomical units (AU) from the Sun. It contains thousands of small icy bodies, including dwarf planets like Pluto, Haumea, and Makemake. The Kuiper Belt provides valuable insights into the early solar system's formation and evolution. It is also the source of many short-period comets. Its discovery in the early 1990s expanded our understanding of the solar system's outer reaches and the diversity of celestial objects beyond the eight major planets.

The Oort Cloud is a theoretical spherical shell of icy objects surrounding our solar system, from about 2,000 to 100,000 astronomical units (AU) from the Sun. It is believed to be the source of long-period comets, with orbits that take thousands to millions of years to complete. These objects are thought to be remnants from the early solar system, ejected by gravitational interactions with giant planets. While the Oort Cloud hasn't been directly observed, its existence is inferred from long-period comet trajectories, helping scientists understand the solar system's formation and outermost regions.

www.ingramcontent.com/pod-product-compliance
Lightning Source LLC
Chambersburg PA
CBHW040029050426
42453CB00002B/61